NATIONAL OCEAN POLICY IMPLEMENTATION PLAN

National Ocean Council

APRIL 2013

National Ocean Council

Council on Environmental Quality

Office of Science and Technology Policy

Department of Agriculture

Department of Commerce

Department of Defense

Department of Energy

Department of Health and Human Services

Department of Homeland Security

Department of the Interior

Department of Justice

Department of Labor

Department of State

Department of Transportation

Environmental Protection Agency

Federal Energy Regulatory Commission

Joint Chiefs of Staff

National Aeronautics and Space Administration

National Oceanic and Atmospheric Administration

National Science Foundation

U.S. Army Corps of Engineers*

U.S. Coast Guard

Office of the Director of National Intelligence

Office of Management and Budget

Office of the Vice President

National Security Staff**

Domestic Policy Council

National Economic Council

*   *Member of the Deputy-level committee*

**   *Includes the National Security Advisor and the Assistant to the President for Homeland Security and Counterterrorism*

EXECUTIVE OFFICE OF THE PRESIDENT
NATIONAL OCEAN COUNCIL
WASHINGTON, D.C. 20503

April, 2013

Dear Colleague:

We are pleased to deliver the *National Ocean Policy Implementation Plan* (Plan), a document that translates the National Ocean Policy into on-the-ground actions that will benefit Americans. The Plan presents specific actions Federal agencies will take to bolster our ocean economy, improve ocean health, support local communities, strengthen our security, and provide better science and information to improve decision-making.

The National Ocean Policy, created by Executive Order 13547 on July 19, 2010, established the National Ocean Council, which consists of 27 Federal agencies, departments, and offices working together to share information and streamline decision-making. The Council developed the Plan over a two-year period with extensive public input from a wide range of stakeholders.

The National Ocean Policy and accompanying Plan will help spur economic growth, empower states and communities, and save taxpayer dollars through better coordination that avoids conflicts. They are examples of common-sense good government that will help Americans sustain and enjoy our ocean resources.

Sincerely,

Nancy H. Sutley
Chair, Council on Environmental Quality
Co-Chair, National Ocean Council

John P. Holdren
Director, Office of Science and Technology Policy
Co-Chair, National Ocean Council

Table of Contents

photos (clockwise from top right): MMS; NOAA; Navy; NOAA; NSF; NOAA

I. Introduction

The ocean, our coasts, and the Great Lakes are among our most treasured resources. They are an integral part of our national identity and our future. A healthy marine environment feeds our Nation, fuels our economy, supports our cultures, provides and creates jobs, gives mobility to our Armed Forces, enables safe movement of goods, and provides places for recreation. Healthy, productive, and resilient oceans, coasts, and Great Lakes contribute significantly to our quality of life.

At the same time, these resources are vulnerable to activities and impacts that diminish their health, productivity, and resilience. Pollution, for example, degrades marine habitats, reduces access to recreational and commercial opportunities, and threatens public health and safety. Habitat loss impacts the stability of marine populations, leading to significant economic and cultural consequences. Overfishing threatens current and future opportunities for recreational and commercial fishing, compromises our national food security, and reduces the ability of marine ecosystems to recover from disturbances. The impacts of climate change, such as sea-level rise, increase the vulnerability of coastal communities to storm damage. Moreover, these problems interact with one another, collectively amplifying their impact on the health of the ocean.

In addition, a growing population of ocean users is increasingly competing for ocean space both for established uses such as fishing, shipping, military activities, and conventional energy development, and for emerging uses such as renewable energy development and aquaculture. This competition creates conflicts between users and presents new challenges for decision-makers. Inefficient government decision-making can compound the problem, hampering economic opportunities and impeding the entrepreneurial, problem-solving efforts of commercial and conservation interests alike.

At the same time, the Nation is encountering new opportunities to improve our understanding of the ocean, how it works, and how we can expand our use of the ocean while maintaining its health and resilience. Advances in research, science, and technology are necessary to help us better understand how marine environments function, and how they influence and are influenced by human activities. Application of this knowledge will inform locally-driven management practices and will improve and maintain the health of the ocean, support employment and new economic opportunities, enhance the Nation's safety and security, and help preserve the ocean as a valuable resource.

Recognizing these challenges and opportunities, and building on the recommendations of two bipartisan commissions, President Obama established the National Policy for the Stewardship of the Ocean, Our Coasts, and the Great Lakes by Executive Order 13547 on July 19, 2010. The National Ocean Policy (Policy) highlights our responsibility to improve and maintain the health of the ocean, coasts, and Great Lakes and recognizes the importance of working with States,[1] tribes,[2] and other partners to tackle key challenges through common sense, science-based solutions. The Policy aims to ensure that our valuable ocean, coastal, and Great Lakes resources will continue to provide a wealth of benefits that support the Nation's well-being, safety, and prosperity.

Fundamentally, the National Ocean Policy coordinates, through establishment of the National Ocean Council, the ocean-related activities of Federal agencies to achieve greater efficiency and effectiveness, with a focus on reduced bureaucracy, improved coordination and integration, and fiscal responsibility. The Policy does not create new regulations, supersede current regulations, or modify any agency's established mission, jurisdiction, or authority. Rather, it helps coordinate the implementation of existing regulations and authorities by all Federal agencies in the interest of more efficient decision-making. The Policy does not redirect congressionally-appropriated funds, or direct agencies to divert funds from existing programs. Instead, it improves interagency collaboration and prioritization to help focus limited resources and use taxpayer dollars more efficiently.

Developed collaboratively by the agencies of the National Ocean Council and based on the work initiated by the Interagency Ocean Policy Task Force in 2009, this National Ocean Policy Implementation Plan (Plan) provides clear direction to Federal agencies and increased specificity to partners and stakeholders.[3] The Plan reflects a commitment to develop and apply the latest science and information, conduct the business of government more efficiently, and collaborate more effectively with State, tribal, and local authorities, marine industries, and other stakeholders. This Plan describes specific actions that translate the goals of the National Ocean Policy into on-the-ground change to address key challenges, streamline Federal operations, save taxpayer dollars, and promote economic growth.

A wide range of stakeholders and partners will benefit from these actions to improve the stewardship and health of the ocean, coasts, and Great Lakes. For example:

- States and tribes will benefit from improved coordination with Federal agencies, better information for decision-making, and support for regional priorities and solutions;

- Recreational fishermen and boaters will benefit from actions that advance conservation programs and help ensure continued access to healthy and productive waters;

- Commercial fishermen will be better equipped to meet our Nation's growing demand for healthy seafood through improved science that supports increased sustainable fishing opportunity;

- The commercial shipping and ports industry will have more accurate charts for safe and efficient navigation;

- The Nation's Armed Forces will benefit from improved coordination with maritime interests to ensure their ability to test and train in order to meet current and emerging national security requirements;

- Offshore energy industries will benefit from better data and information to identify potential development sites, more efficient leasing and permitting processes, and planning that facilitates safe access, safe operations, and reduced conflicts with other uses;

- The aquaculture industry will benefit from streamlined Federal permitting and coordinated research efforts to support sustainable aquaculture;

- Coastal communities will build resilience to extreme events and sustain more coastal job opportunities; and

- Beach-goers, birders, conservationists, and others will benefit from healthier coastal and ocean habitats and ecosystems.

This Plan presents a common-sense, science-based approach to achieve these benefits through resource management that considers entire ecosystems. The goal of ecosystem-based management supported by this Plan is to maintain a healthy, productive, and resilient ocean that can continue to provide the benefits and resources humans want and need. Achieving this goal will require both a sound scientific foundation and a commitment to management practices that are adaptable to changing conditions and responsive to new challenges and opportunities that emerge. Working together, resource managers, ocean users, and other stakeholders can develop and apply ecosystem-based management incrementally, by learning and sharing effective practices as knowledge and experience increase.

Importantly, this Plan was informed by thoughtful input from national, regional, and local stakeholders from all marine sectors; tribal, State, and local governments; private sector partners, academic scientists, and the general public. It reflects careful consideration of extensive public comments, particularly those that relate to the importance of incremental change, pilot projects, support for local and regional capacity and self-determination, and the fundamental need for more and better information.

The Implementation Plan better aligns multiple agency priorities and activities to promote greater synergies and efficiencies in Federal spending. Given today's constrained fiscal climate and recognizing uncertainty in the budget and appropriations processes, completion of every action within the identified timeframes will depend upon the availability of funds and resources.

In that vein, this Plan is intended to be a living document. It is designed to be adaptive to new information or changing conditions, and will be updated periodically as progress is made, lessons are learned, new activities are planned, and as the Nation continually strives to improve the stewardship of the ocean, coasts, and Great Lakes for the benefit of current and future generations.

Organization of the Document

This Implementation Plan describes, under the following sections, how specific actions to implement the Policy will benefit: (1) The Ocean Economy, (2) Safety and Security, and (3) Coastal and Ocean Resilience by supporting (4) Local Choices, and providing foundational (5) Science and Information. Subsections describe specific outcomes that advance those benefits and the types of actions Federal agencies will take to achieve them. Specific planned actions are described in the appendix containing Implementation Actions. Many of these actions will produce benefits in the short-term that respond to immediate needs of communities, ocean stakeholders, and the public. Others create building blocks to support key outcomes in the medium- to long-term. The actions in this Implementation Plan are grounded in the National Priority Objectives of the Policy. They encompass efforts previously identified under these objectives as those that will move our Nation ahead toward resolving the most pressing challenges facing the ocean, our coasts, and the Great Lakes, and benefitting the people, communities, and businesses that rely on them.

Many of the actions support multiple outcomes, reflecting the common-sense value of focusing and coordinating the work of Federal agencies to provide products and services that benefit all Americans. In particular, a number of science and information-based actions that advance observing systems, mapping and charting, and other information tools benefit many different users at the national, regional, and local level. Those and other actions are therefore discussed in more than one section to explain how they advance each policy objective.

photo: NOAA

II. The Ocean Economy

The ocean, our coasts, and the Great Lakes are among our Nation's most valuable resources and strongest economic drivers. In 2010, maritime economic activities such as shipping, marine construction, energy development, commercial fishing, recreational fishing and boating, aquaculture, and tourism contributed $258 billion in GDP to the national economy and supported 2.8 million jobs.[4] Because so many people live near the coast, in 2010, 41 percent of our Nation's Gross Domestic Product (GDP), or $6 trillion, was generated in the shoreline counties of the United States and territories, including the Great Lakes.[5] These coastal counties supported approximately 44 million jobs and $2.4 trillion in wages.[6] The value of the ocean to Americans—for commerce, energy, recreation, food, culture, and national security—provides the foundation for our quality of life now and for future generations.

As a maritime Nation, we are challenged to maintain and enhance the economic benefits that a healthy and productive ocean provides. The declining health of ocean, coastal, and Great Lakes ecosystems threatens their ability to provide the products and services on which much of our economy depends. For example, marine and aquatic invasive species cost our economy billions of dollars each year in damage to fisheries, tourism, and coastal infrastructure.[7] Another example indicates that coral bleaching has cost the United States an estimated $4.8 billion over the past 50 years, affecting tourism and fishing, and increasing the vulnerability of coastal areas to storm damage.[8] The proliferation of marine debris along our coasts has significant economic impacts across a number of marine sectors, including tourism, recreation, and fisheries.[9]

Government inefficiencies can add to these problems. For example, of the seafood consumed in the U.S. in 2011, an estimated 91 percent (by value) was imported with half of that coming from foreign aquaculture.[10] In 2011, the U.S. trade deficit in seafood was $11.2 billion, a number that grows annually.[11] Government inefficiency in the siting, permitting and approval processes for aquaculture may be hindering the domestic aquaculture industry's growth.[12] Beyond threatening jobs and economic stability, poor coordination and ineffective planning can cause increased delays, conflicts, and costs among the growing number of ocean users.

This Plan responds to such challenges by focusing and coordinating action among Federal agencies under their existing authorizations and budgets, and by providing the tools we need to ensure a robust, sustainable ocean economy. It also promotes better science and information to support economic growth, more efficient permitting and decision-making, and healthier and more resilient marine ecosystems that will continue to support jobs, local economies, and a skilled and diverse ocean workforce.

A healthy marine environment provides significant economic benefits. For example, millions of Americans experience the ocean, coasts, and Great Lakes each year through recreational fishing and boating, which is a major contributor to the national economy. In 2010, marine tourism and recreation accounted for 70 percent of the jobs produced by the total ocean economy—1.9 million American jobs in total.[13] As such, maintaining healthy, productive waters and access to them for recreation and other activities is critically important to sustaining the benefits that so many Americans enjoy. The recreational fishing and boating communities directly contribute to and help fund (through excise taxes and license sales) many marine conservation, State wildlife and fishery programs, and other initiatives that provide further benefits through vehicles such as the Sport Fish Restoration and Boating Trust Fund. These are just some examples of the value provided by healthy marine waters.

The following actions will support existing and new marine industries, maintain and enhance the vitality of coastal communities and regions, and preserve the marine ecosystems that support our quality of life.

Supporting Economic Growth

Businesses, communities, and governments that rely on ocean resources need high-quality scientific information and data. Greater access to high-quality data and information will enable maritime industries, resource managers, and decision makers at all levels of government to make responsible and effective decisions. Federal agencies will take the following actions that strengthen the national economy through enhanced accessibility to data and information and robust, sustained observing systems.

- **Advance our mapping and charting capabilities and products to support a range of economic activities.** To sustain the flow of the trillions of dollars of goods that pass through our ports and the many businesses that rely on the ocean, our coasts, and Great Lakes, agencies will coordinate to produce better mapping and charting products, which serve to preserve, protect, and expand our Nation's maritime economic activities. Improved mapping, charting, and associated products will enhance the efficiency of maritime commerce through safer navigation and better accident-avoidance, and updated hydrographic charts and seafloor maps will support marine industries such as offshore energy. These products will also provide coastal communities with better elevation and bathymetric data to plan for and mitigate the adverse economic impacts of disasters.

- **Provide greater accessibility to data and information to support commercial markets and industries, such as commercial fishing, maritime transportation, aquaculture, and offshore energy.** Agencies will take a series of actions to facilitate the availability of relevant ocean data to provide easier access to information for research, planning, and decision support. Further, agencies will utilize public input, local and traditional knowledge, and scientific information to help identify and communicate the economic value of ecosystem services, such as healthy and productive wetlands that support spawning, breeding, and feeding of commercially and recreationally important fish species. This information can help decision makers consider the value of these services when evaluating actions that may impact the economy.

- **Sustain and further develop observing systems for the economic benefit of maritime commerce and marine industry.** Federal agencies will support the development and maintenance of ocean observing systems. Real-time information on waterway conditions from ocean, coastal, and Great Lakes observing systems such as the Physical Oceanographic Real-Time System directly supports the daily operations and efficiency of maritime commerce nationwide, as well as local and regional businesses that rely on the marine environment. Continued development of Federal ocean observing programs will stimulate private sector ocean technology development and provide a rigorous test-bed for new innovations.

Promoting Jobs

Ocean industries are a major employer. In 2010, U.S. commercial ports supported more than 13 million jobs.[14] Similarly, in 2011, commercial fisheries supported 1.2 million jobs and $5.3 billion in commercial fish landings, and marine recreational fisheries supported 455,000 jobs.[15] As of March 2012, energy and minerals production from offshore areas accounted for about $121 billion in economic contributions to the U.S economy and supported about 734,500 American jobs.[16] Offshore wind energy has the potential to directly support 20.7 jobs for every megawatt-hour generated. Installing 54 gigawatts of offshore wind capacity in U.S. waters would create more than 43,000 permanent operations and maintenance jobs.[17] There is significant potential along the Nation's shorelines to create a large number of coastal restoration jobs that recover degraded habitats and restore the fisheries and recreational opportunities they provide. For every million dollars invested, coastal restoration creates between 17 and 30 new jobs for coastal regions—regions that provide key habitat for more than 70 percent of the commercial and recreational fish catch.[18] Marine aquaculture in the U.S. has a farm-gate value of $320 million[19] and supports up to 35,000 jobs.[20] Supporting the growth of sustainable marine aquaculture through the National Shellfish Initiative and building on existing efforts such as the Gulf of Mexico Fishery Management Council's Aquaculture Plan has the potential to provide additional jobs.

The following actions by Federal agencies will help maintain existing jobs and promote job growth in coastal and marine-related sectors by improving regulatory efficiency, reversing environmental impacts that hinder economic opportunity, and providing information that supports actions to maximize the economic value of our natural resources. The goal of these actions is to enhance both immediate and long-term potentials for job creation.

- **Increase efficiencies in decision-making by improving permitting processes and coordinating agency participation in planning and approval processes.** A key goal of the Policy is to improve efficiency across Federal agencies, including permitting, planning, and approval processes to save time and money for ocean-based industries and decision makers at all levels of government while protecting health, safety, and the environment. Interagency work already in progress includes more efficient permitting of shellfish aquaculture activities, which will help produce additional domestic seafood and jobs and provide a template for similar action to support other marine commercial sectors. Through pilot projects developed in collaboration with relevant stakeholders, Federal agencies will identify opportunities to streamline processes and reduce duplicative efforts while ensuring appropriate environmental and other required safeguards.

- **Provide jobs and economic value by protecting and restoring coastal wetlands, coral reefs, and other natural systems.** Restoration activities provide direct economic opportunities, and healthy natural systems support jobs in industries such as tourism, recreation, and commercial fishing. Agencies will coordinate to protect, restore, and enhance wetlands, coral reefs, and other high-priority ocean, coastal, and Great Lakes habitats. Agencies will also work through the already established a National Shellfish Initiative with commercial and restoration aquaculture communities to identify ways to both responsibly maximize the commercial value of shellfish aquaculture and achieve environmental benefits such as nutrient filtration and fish habitat.

- **Prevent lost employment opportunities and economic losses associated with environmental degradation.** Hypoxia and harmful algal blooms have significant adverse economic, public health-related, and ecological consequences. Invasive species are a major challenge that results in economic losses to local communities and industries, costing the Nation more than $120 billion annually.[21] Federal agencies will take steps to prevent and reverse widespread economic impacts caused by hypoxia, harmful algal blooms, invasive species, and other threats to healthy systems. They will take action to strengthen the monitoring, science, data access, modeling, and forecasting of hypoxia and harmful algal blooms to provide decision makers with the necessary information to minimize and mitigate harmful impacts on coastal economies. Federal agencies will take actions to improve our ability to detect and reduce invasive species in coastal and ocean habitats to protect commercial and recreational fish stocks, help sustain the jobs and industries that depend upon healthy coastal aquatic ecosystems, and save millions of dollars in lost revenue and avoided infrastructure damage.

Developing a Skilled Ocean Workforce

A diverse workforce with interdisciplinary skills and training is needed to maintain the Nation's place as a world leader in ocean science and to ensure informed management and use of ocean, coastal, and Great Lakes resources. Agencies will coordinate to build the science, technology, engineering, and mathematics (STEM) and managerial workforce capacity needed to ensure that management of and research on ocean, coastal, and Great Lakes ecosystems are of the highest quality possible.

- **Develop human capacity and the skilled workforce necessary to conduct ocean research and manage ocean resources.** Agencies will coordinate to ensure that educational programs include diverse student groups and that a highly competent workforce is developed. Agency actions will result in more students, particularly from underrepresented groups at the undergraduate and graduate level, pursuing academic fields related to ocean, coastal, and Great Lakes science and management. This will support the Nation's leadership in ocean research and development and the application of best management practices. For example, agencies will use existing education and training resources to provide scholarship, fellowship, and internship opportunities that leverage existing Federal investments in ocean research, marine laboratories, and natural sciences to provide opportunities for education and training. Agencies will also contribute to periodic ocean-focused academic competitions for middle and high school students that have a positive impact on ocean-related career paths.

photo: UNH/NOAA

III. Safety and Security

The ocean, our coasts, and Great Lakes are critically important to the Nation's safety and security. Safe, secure, and productive access to, and use of, our maritime domain are essential to maintaining military strength, a strong economy, and a high quality of life for all Americans. Marine waters comprise the physical boundaries of our Nation, support the mobility and training of our Armed Forces, and provide an economically vital foundation for energy, commerce, tourism and recreation, commercial and recreational fishing and boating, and other industries. For many Native and tribal communities and coastal residents, these waters directly sustain life and cultures. It is fundamentally in our Nation's best interest to better understand, protect, and sustain these waters.

Industry, government, academia, and the public conduct numerous activities on our coastlines and in the ocean and Great Lakes for a variety of purposes. By improving effective coordination and situational awareness, these comingled activities will take into consideration the safety and security of our people, property, and the health of the marine environment. Federal agencies will work together to improve our overall awareness of the maritime domain, be responsible stewards of the marine environment, and enhance the safety and security of our ports and waterways.

International cooperation is equally important. United States accession to the Convention on the Law of the Sea (Convention) is critical to protecting our navigational rights and freedoms, both for military vessels and for civilian vessels and their cargoes, and to advancing our economic interests in the ocean. The Convention accords to the United States extensive offshore resource rights, including exclusive rights to natural resources such as oil, gas, and fish, out to 200 nautical miles from shore, and additional rights to seabed resources, including oil and gas, beyond 200 nautical miles in several large areas. Accession to

★ 10 ★

the Convention also means that the United States would have the opportunity to place U.S. nominees/designees on various Convention bodies, including those developing the rules governing mineral resources in the deep seabed, and those making recommendations regarding Parties' submissions on the continental shelf beyond 200 nautical miles. Joining the Convention will advance our national interests by protecting and enhancing our access to the ocean and important natural resources.

Improving Maritime Domain Awareness

A solid understanding of the wide range of activities, infrastructure, and environmental conditions that occur in the ocean, coastal, and Great Lakes ecosystems enables informed reactions and responses to events that occur in those waters. Maritime domain awareness is achieved by efficiently collecting and sharing information and by improving the Nation's infrastructure for ocean observing and remote sensing systems. It is also important to focus on greater collaboration with the international community to enable better sharing of information, expertise, and knowledge with other nations.

- **Enhance remote sensing systems for ocean observations to support maritime domain awareness.** Federal agencies will optimize use of enhanced remote sensing systems for ocean observations to improve awareness of real-time oceanographic, meteorological, and ecological conditions in the maritime domain. An integrated system of remote sensing assets designed for ocean observations will assist decision-makers by providing a more complete picture of the marine environment.

- **Engage internationally to exchange information, expertise, and knowledge about policy issues in the maritime domain.** The United States will collaborate with international organizations and bodies, such as the International Maritime Organization and Intergovernmental Oceanographic Commission, and with other nations, in exchanging information, expertise, and knowledge to address high-priority ocean policy issues. These efforts will improve awareness of activities in the maritime domain, especially among those nations sharing a maritime border with the United States, and enhance our ability to address high-priority ocean policy issues efficiently and effectively.

Providing Maritime Safety and Security in a Changing Arctic

The Arctic is rapidly changing. One of the most dramatic changes is the decrease in sea ice, which is likely to increase vessel traffic in the U.S. Arctic. Commercial vessels may capitalize on more expeditious routes, cruise ships and recreational vessels are expected to bring more tourists to the region, fishing grounds are shifting, and oil and gas companies are moving forward with exploration activities and obtaining leases to drill into the Arctic seabed. This brings a need for improvements to our Arctic communication systems and environmental response management capabilities; our ability to observe and forecast sea ice; and the accuracy of maps and charts of the region. Our maritime safety and security in the Arctic hinge upon these actions.

- **Enhance communication systems in the Arctic to improve our capability to prevent and respond to maritime incidents and environmental impacts.** Federal agencies will improve Arctic communication systems by advancing both technical capabilities and partnerships.

Agencies will strengthen existing communication systems to allow vessels, aircraft, and shore stations to effectively communicate with each other and to receive information such as real-time weather and sea ice forecasts that will significantly decrease the risk of loss of life at sea or damage to property or the marine environment. Agencies will partner with each other, Native communities, industry, and other countries as appropriate to identify user needs and existing capabilities prior to building new communication systems.

- **Improve Arctic environmental incident prevention and response to ensure coordinated agency action, minimize the likelihood of disasters, and expedite response activities.** Increased Arctic vessel traffic brings increased risks of collisions, groundings, and other serious marine incidents that can lead to loss of life and property and damage the marine environment. A coordinated and prepared all-hazards response-management system will mitigate the impacts of marine-pollution events on fragile Arctic communities and ecosystems. To improve responses, Federal agencies will conduct joint spill-response workshops and exercises, develop and implement response coordination and decision-support tools like the Arctic Environmental Response Management Application, and improve spill prevention, containment, and response infrastructure, plans, and technology for use in ice-covered waters.

- **Improve Arctic sea ice forecasting to support safety at sea.** Sea ice forecasting is one of the most urgent and timely issues in the Arctic region. To ensure the best tactical and long-term ice forecasts are available for safe operations and planning, Federal agencies will work together to better quantify the rates of sea ice melt and regrowth, understand shifting patterns of distribution of ice, develop better maps of the ice edge, expand participation in the sea ice observation program, and coordinate with international partners to enable better model-based forecasting over larger areas. Improved observations will contribute to improved forecasts, which will better inform Arctic maritime safety and security activities.

- **Improve Arctic mapping and charting for safe navigation and more accurate positioning.** Advancements in hydrographic charting will enhance the safety of navigation in the Arctic region by reducing the risk of damaging maritime incidents. Federal agencies will update nautical charts and establish priorities, in concert with Native communities and stakeholders, for shoreline and hydrographic surveying activities. Further, mapping gravity data over the State of Alaska will help correct meters-level errors in Arctic positioning. Such efforts will support U.S. Navy and U.S. Coast Guard operations and help ensure the safety and security of all mariners in the Arctic.

Enhancing the Safety and Security of Ports and Waterways

The safety and security of our people, property, and the marine environment and the viability of maritime commerce rely on safe, efficient, and secure navigation and waterways management systems. This includes effective planning for and response to emerging threats to our ports and harbors from illegal human activities, climate change, and extreme weather events or other natural disasters. Federal agencies will conduct several actions that leverage existing resources in a coordinated manner to ensure the safety and security of all those who make a living from, enjoy through recreation, and rely on the health and vitality of our ports and waterways.

- **Conduct Waterway Analysis and Management System assessments and Port Access Route Studies to support decisions on waterways management and other navigational priorities.** The safe and secure navigation of commercial, recreational, and government vessels in and out of our Nation's ports depends on accurate and timely assessments of our waterways. Federal agencies will evaluate the performance of our waterways management systems on a consistent basis to identify the improvements needed to ensure the safety and security of our maritime public, the economic vitality of our ports, and the integrity of our marine ecosystems.

- **Assess the vulnerability of our ports and waterways to sea-level rise and extreme weather events or other natural disasters and enable actions that more effectively reduce risks and impacts.** The Nation's ports and waterways infrastructure support many economic, safety, and security activities. A better understanding of the potential impacts of climate change on our ports and waterways will prepare us to respond and adapt accordingly in order to preserve critical assets. Vulnerability assessments are critical to understanding how extreme weather, sea-level rise, and other manifestations of climate change may affect our coastlines. Federal agencies will conduct such assessments, in collaboration with State, tribal, local, and regional efforts. Federal agencies will develop and disseminate methods, best practices, and standards for assessing the resilience of natural resources, populations, and infrastructure in a changing climate.

- **Advance ocean observing systems to further enhance search and rescue operations and spill response in our ports and waterways.** Ocean observing systems provide real-time and near real-time oceanographic, meteorological, and ecological data, which feed into search and rescue and oil spill trajectory models. The reliability, quality, and resolution of ocean observing system data have a direct impact on the model output, which influences operational decisions for search and rescue and oil and hazardous substance spill response. Advancing the capabilities and reliability of our ocean observing system infrastructure will further protect life, property, and the marine environment in our economically vital ports and waterways.

photo: USFWS

IV. Coastal and Ocean Resilience

The health and integrity of coastal habitats—such as coral reefs, wetlands, mangroves, salt marshes, and sea grass beds—are key to sustaining our Nation's valuable coastal and ocean ecosystems and the wealth of benefits they provide to us. Outdoor enthusiasts across the Nation access coastal habitats for fishing, boating, birding, and hiking; communities depend on coastal wetlands as buffers against hurricanes; divers and snorkelers enjoy the beauty of coral reefs; and commercial fishermen provide fresh seafood for our tables. Coastal habitats provide spawning grounds, nurseries, shelter, and food for finfish, shellfish, migratory birds, and waterfowl. They protect coastal communities, homes, infrastructure, and businesses against damage from erosion and flooding, they support hundreds of thousands of jobs, they improve water quality by filtering and detoxifying runoff, they dampen the outbreak of pests and pathogens, they capture and store carbon, and they yield compounds for life-saving medicines. Healthy watersheds and coasts sustain our Nation by providing abundant clean water to coastal communities, residents, businesses, industries, and ecosystems.

Degradation of coastal habitats and ecosystems diminishes their health and their ability to provide environmental, economic, and societal services to the Nation. Our Nation lost nearly 60,000 acres of coastal wetlands each year between 1998 and 2004.[22] Roughly half of the coral reefs under U.S. jurisdiction are in "poor" or "fair" condition because of ocean warming, disease, and human activities.[23] Habitats are being altered by invasive species that threaten native aquatic life and cost billions of dollars per year in natural and infrastructure damage.[24] Trash in the Nation's waterways injures and kills wildlife, degrades habitats, interferes with navigation, threatens public health and safety, and creates additional costs for shipping, fishing, tourism, and coastal communities. Pollution from a variety of sources affects our streams, rivers, estuaries, and coasts, and is the leading cause of water quality problems in the United States. Such pollution represents a major cause of rapidly declining ocean and coastal ecosystem health.

These threats are exacerbated by the environmental impacts of climate change and ocean acidification and the resulting shifts in wildlife populations and abundance. Sea-level rise, increased severe storm events, changing ocean temperature, and saltwater intrusion present serious and growing threats to low-lying coastal communities through the destruction of infrastructure, flood inundation, loss of arable land, and the potential displacement of millions of people.[25] Climate change is also predicted to alter water levels of the Great Lakes, thereby changing water cycles and supply, habitats, and economic uses of the Lakes. Ocean acidification, caused by increased levels of carbon dioxide that make marine waters more acidic, can harm the growth of plants and animals, including recreationally and commercially important fish and shellfish. Marine industries such as shellfish aquaculture, and the jobs and communities they support, face increased impacts from the changing chemistry of our marine waters.

Federal agencies will work together to support the various national, State, tribal, and local efforts to prepare for, respond to, and mitigate or avoid the degradation and loss of ocean and coastal habitats, water quality, and ecosystems through improved capabilities, proactive stewardship, strengthened research, and enhanced collaboration. Agencies will also enable and support efforts to understand, minimize, and adapt to the impacts of climate change, ocean acidification, sea-level rise, and extreme weather events, strengthening the resilience of coastal communities.

Reducing Adverse Conditions

Through National Ocean Policy actions, thousands of acres of wetlands and priority habitat will be protected, restored, or enhanced. Our Nation's coral reefs will be improved by better coordinating existing authorities and implementing projects to prevent or mitigate harmful impacts. Actions to support partnerships and efforts to locate, monitor, control, and eradicate invasive species will protect native aquatic populations and their habitats. Collaborative watershed restoration efforts are important to the overall success of coastal and marine habitat conservation. Restoration efforts in the Gulf Coast, Mississippi River Basin, and Great Lakes, and for Pacific Northwest salmon are excellent examples of collaborative, voluntary upland watershed conservation and restoration.

- **Reduce coastal wetland loss.** Federal agencies will work together and in cooperation with States and tribes to identify the underlying causes of wetland loss in coastal watersheds, and opportunities to more effectively protect and restore the important functions and values they provide. Agencies will conduct pilot studies to identify the most common underlying factors responsible for coastal wetland loss and the most successful tools for addressing it. These actions will complement ongoing State, local, and tribal government projects seeking to protect and restore coastal wetland ecosystems such as the Gulf Coast Ecosystem Restoration Council and the South Florida Ecosystem Restoration Task Force.

- **Protect, conserve and restore coastal and ocean habitats.** Agencies will coordinate to use and provide scientifically sound, ecosystem-based approaches to achieving healthy coastal and ocean habitats. For example, working through the U.S. Coral Reef Task Force, agencies will coordinate to address key threats to coral reef ecosystems, including impacts from land-based sources of pollution, climate change, ocean acidification, planned activities (authorized activities), and unplanned activities (such as vessel groundings and spills).

- **Locate, control, prevent, and eradicate invasive species populations.** Federal agencies will improve our ability to prevent and reduce impacts from invasive species, focusing on early detection and response, to protect ecologically, commercially, recreationally, and culturally, important marine species and their habitats.

- **Improve and preserve our Nation's coastal and estuarine water quality to provide clean water for healthier waterways, communities, and ecosystems.** Through more effective use of voluntary programs, partnerships, and pilot projects, agencies will work to reduce excessive nutrients, sediments, and other pollutants. Agencies will also help protect, conserve, and maintain high-quality coastal waters by identifying priority areas for water quality monitoring and assessment and providing financial assistance to private landowners seeking to apply voluntary conservation practices. Other actions will reduce the impacts of hypoxia and harmful algal blooms faced by many coastal and inland States.

Preparing for Change

Agencies will take a number of actions to improve the resilience of coastal communities and enhance their ability to adapt to the impacts from climate change, extreme weather events, and ocean acidification. Agencies will develop estimates for global mean sea-level rise and make available coastal inundation and sea-level change visualizations and decision-support tools relevant to regional, State, tribal, and local decision-makers. They will offer tools and training courses on how to design and implement vulnerability assessments and develop a national assessment of coastal and ocean vulnerability to both climate change and ocean acidification.

Actions will be conducted in coordination with other Federal climate change and ocean acidification programs and strategies, including the National Action Plan for Managing Freshwater Resources in a Changing Climate, the National Fish, Wildlife and Plants Climate Adaptation Strategy, the Strategic Plan for Federal Research and Monitoring of Ocean Acidification, the U.S. Global Change Research Program, the National Climate Assessment, and the Interagency Climate Change Adaptation Task Force. To the extent appropriate, these actions will also be coordinated with and guide relevant Federal Emergency Management Agency efforts such as national preparedness, disaster response and recovery, and flood hazard map development.

- **Strengthen and integrate observations into a coordinated network of sentinel sites to enhance the Nation's ability to provide early warnings, risk assessments, and forecasts for impacts.** Federal agencies will strengthen and integrate observations from the Nation's protected areas, research sites, and observing systems into a coordinated network of climate sentinel sites. This is an efficient and effective way to provide decision-makers with the information they need to reduce risks and increase resilience of ocean and coastal environments and communities in a changing climate.

- **Determine the impacts of interacting stressors on ecological systems, economies, and communities.** Agencies will develop an integrated research agenda to help address gaps in our current understanding of impacts due to multiple, interacting factors, and build a foundation for the development of models, tools, and services to better inform future planning and decisions and improve implementation of existing policies. This integrated, interdisciplinary agenda will provide information for better forecasts of changes in ecological, economic, and social systems due to climate change and ocean acidification, and improved effectiveness of adaptation actions, with the goal of reducing risks and negative impacts on communities. For example, enhanced sea-level rise projections can inform the development of flood hazard maps.

- **Assess the vulnerability of coastal communities and ocean environments to climate change and ocean acidification and, in partnership with tribes, coastal communities and States, design and implement adaptation strategies to reduce vulnerabilities.** Agencies will develop methods, best practices, and guidance for assessing the vulnerability and resiliency of resources to a changing climate, building off existing efforts such as the National Climate Assessment. These tools will allow decision makers to assess local vulnerability, avoid actions that increase vulnerability of human communities or degrade natural resources, and take actions that increase resilience of both natural systems and communities. Agencies will also strengthen the institutions, mechanisms, and capacities for systematically enhancing resilience to hazards and incorporating adaptation strategies for coastal and ocean species and habitats into future planning, management processes, and infrastructure investments.

Recovering and Sustaining Ocean Health

Agencies will take a number of actions to significantly improve our Nation's capacity to address the long-term challenges and impacts of natural and human-caused environmental changes. These actions will strengthen collaboration through scientifically sound ecosystem-based management.

Effective management of activities that affect ocean health requires considering several inter-connected functions of ocean ecosystems, the resources they provide, and how human activities impact both the ecosystems themselves, and the communities that depend on them. Most previous management approaches have focused on a single resource or issue and designed solutions focused on that resource or issue alone. For example, the conventional approach to managing fisheries has been to focus on a single species and work to ensure its availability, primarily by limiting how many of them can be caught. Integrated, ecosystem-based management goes further and supports the goal of having a greater abundance, distribution, and diversity of fish, more jobs, and thriving fishing communities by also addressing the food sources and ecosystems that fish need to grow and the factors that affect them. By understanding those connections, managers can make decisions that support all components of the system, so there can be more fish overall. Federal fishery managers are already applying a more integrated management approach, but it does not include consideration of non-fishery factors, such as water quality, that affect fisheries.

Together, the following actions will provide a lasting foundation for enhancing the many vital benefits our Nation derives from healthy ocean, coastal, and Great Lakes ecosystems.

- **Establish a framework for collaboration and a shared set of goals to promote ecosystem-based management.** Agencies will increase their collaboration with other levels of government, experts, practitioners, and stakeholders to enhance the efficiency, consistency, and transparency of their development and implementation of ecosystem approaches to management based on existing statutes and regulations. Agencies will develop principles, goals, and performance measures that support the development of integrated ecosystem-based management.

- **Improve coastal and estuarine restoration efforts through better monitoring, coordination, and planning.** Monitoring restoration efforts provides important data and information to improve the science of restoration and track the societal benefits of restoration activities, such as increased fish populations and enhanced protection of coastlines from storms. Federal agencies that fund and implement coastal and estuarine habitat restoration projects will evaluate and track these efforts to ensure that they are efficient and effective.

- **Improve the Nation's preparedness for, and response to, environmental hazards through better forecasts, increased and more integrated monitoring, and strengthened preparedness.** Agencies will establish a Health Early Warning System that alerts public health officials and managers to marine-related threats to human and ecosystem health from diseases, toxins, and pathogens. To enhance our Nation's food safety and security, other actions will augment contaminant monitoring and disease surveillance programs in a target region, and develop new, rapid assessment methods to detect contaminants and spoilage in seafood. Further actions will also reduce the negative impacts of trash and marine debris by enhancing non-regulatory prevention, reduction, and removal methods through methods such as community-based grants. In addition, agencies will develop and implement a coordinated response management system to better protect Arctic communities and ecosystems from potential oil spills and other pollution events.

- **Protect significant natural and cultural marine and Great Lakes areas and sufficient habitat to ensure maintenance of ecosystem processes.** Identifying ecologically important and culturally significant areas for focused protection or management supports the long-term sustainability of ocean resources. Several Federal agencies have processes by which to identify important marine areas for management or protection under existing authorities. Agencies will address, with input from State, tribal, regional, local, and stakeholder interests, the protection of essential fish habitat and support reactivation of the National Marine Sanctuary Site Evaluation List. This List is a public process tool for evaluating marine areas that may be considered for national marine sanctuaries in a transparent and public way.

photo: Dawn Standifur

V. Local Choices

Throughout the U.S., there are myriad tribal, State, regional, and local efforts to support and grow marine economies, protect and conserve the environment that supports quality of life, and sustain unique social and cultural identities. Priorities, however, vary across regions, as do the ways in which different regional actors choose to address them.

All regions share an interest in growing their economies and providing jobs that support strong communities, which they address through a diverse and often unique array of marine uses. For example, their interests range from conventional to renewable energy, they have different commercial and recreational fisheries, and they offer distinct tourism and recreational activities. They also have different priorities for environmental protection and the use of ocean resources. In the Pacific and Caribbean, coral reef ecosystem conservation is a focus area, while in the Pacific Northwest, addressing the impacts of ocean acidification on local shellfish growers is a top priority. In the Gulf of Mexico, efforts are underway to minimize the impacts of harmful algal blooms on human health, while in the Great Lakes, States are working to control invasive species to minimize the damage they cause to commerce, municipal infrastructure, and the Great Lakes ecosystem.

In Arctic communities, adapting to the impacts of climate change is a regional priority, while Chesapeake Bay communities focus on approaches to improve water quality. Regardless of the specific issues being addressed, communities and stakeholders need more and better information and coordinated and responsive Federal agency actions that address locally relevant issues. Actions under the National Ocean Policy provide tools and services that support and build on action at local, State, tribal, and regional scales. These will strengthen partnerships across all levels of government and with regional and local stakeholders and communities.

Providing Tools for Regional Action

Science and data provide the building blocks for information and tools to support tribal, State, local, and regional action. Efficient access to observations and information is improving our ability to understand and predict ecosystem events—such as a loss or change in habitat or coral bleaching—as well as long-term planning and decision-making. Pilot projects focused on ecosystem-based management allow scientists, managers, and stakeholders to account for and address the many factors that affect how ecosystems work, at a manageable scale and in the context of relevant issues. More efficient discovery of, and access to, information improves the ability of tribal, State, local, and regional planners to understand, predict, and prevent or mitigate events. Assessing vulnerability is yet another crucial step in preparing for and responding to the impacts of climate change, ocean acidification, and extreme weather on ocean environments and coastal communities.

- **Identify and implement pilot projects that use an ecosystem-based approach to partnering in the stewardship of ocean and coastal resources.** In collaboration with local, regional, and tribal practitioners, agencies will identify and conduct pilot projects that incorporate best practices for ecosystem-based management, test on-the-ground effectiveness of decision-support tools, and demonstrate the practical utility of ecosystem-based approaches. Pilot projects will determine what additional data, tools, and training are required, identify how the collaborative and scientific frameworks may need to be altered to be more useful, and enable decision-makers and managers to understand how ecosystem-based management can be most effectively implemented at regional scales relevant to address specific resource management objectives.

- **Assess the vulnerability of communities and ocean environments to climate change and ocean acidification and support and implement adaptation strategies to promote informed decisions.** Agencies will develop best practices and guidance for assessing the vulnerability and resilience of communities, infrastructure, and resources to a changing climate and ocean acidification, and will develop and promote adaptation tools and strategies to help coastal communities address these risks. These tools will enable decision-makers at all levels of government to assess local vulnerability, inform near-term and long-term investments, and avoid actions that increase vulnerability.

- **Expand and improve discovery of and access to non-classified Federal data and decision-support tools, including ocean and coastal mapping products, to support local, tribal, State, and regional decision-making.** Not all existing Federal data are easily accessible or in a useable format for regional decision-making and planning purposes. Agencies will coordinate to make unrestricted Federal data publicly available in a standards-based format through a national data portal (ocean.data.gov). This central portal for planning-related ocean, coastal, and Great Lakes data will allow for easy discovery and access to data and derived products which support the further development of new and/or improved decision-support tools for planners at all levels of government.

Strengthening Regional Partnerships

Federal agencies will work to strengthen and leverage existing regional partnerships and to build new ones. Agencies will honor the government-to-government relationship, trust obligations, and consultation responsibilities of the Federal government with Federally-recognized tribes and expand partnerships with tribes and Native communities. Agencies will also partner with and assist States in advancing the network of regional alliances to protect ocean, coastal, and Great Lakes health. Partnerships with local governments and private interests are also needed to leverage limited resources.

Existing regional ocean and Great Lakes partnerships are voluntary, usually multi-state forums established by State Governors that identify shared priorities and take critical action on a range of issues relevant to their region. They have different structures and employ varied methods and approaches to enhance the ecological and economic health of the region. These efforts involve non-governmental stakeholders and multiple State and Federal agencies involved in coastal and ocean management. Federal agency actions will increase communication among ocean sectors, streamline processes, leverage resources, and enhance coordination among all levels of government.

- **Support regional priorities and enhance regional partnerships' ability to address issues of regional importance.** Federal agencies will enhance on-the-ground progress by supporting regional priorities such as data collection and analysis, and by improving coordination among Federal offices based in regions. Agencies will identify opportunities to leverage resources and partner on the continued development and organization of regional alliances and existing partnerships. This will include data collection and analysis needed to advance regional efforts, compile resources available to enhance accomplishment of mutual regional goals, and identify and distribute best management practices that are broadly applicable for all regional ocean and coastal entities (for example, how to effectively engage stakeholders, develop partnerships, identify priorities, develop regional action plans, and measure success).

- **Support engagement of interested tribal authorities and use of tribal information.** Agencies will work with interested tribal governments to support tribal involvement in priority-setting and planning for each region, including the integration of traditional ecological knowledge and scientific data collected by indigenous groups. Agency engagement of and coordination with tribes will ensure that tribal interests, lands, treaty and other reserved rights, and co-management agreements are appropriately considered and included in each region.

Supporting Regional Priorities

Marine planning is a science-based tool that regions can use to address specific ocean management challenges and advance their economic development and conservation objectives. Marine planning will support regional actions and decision-making and address regionally determined priorities, based on the needs, interests, and capacity of a given region. Just as Federal agencies work with States, tribes, local governments, and users of forests and grasslands, among other areas, marine planning will provide a more coordinated and responsive Federal presence and the opportunity for all coastal and ocean interests in a region to share information and coordinate activities. This will promote more efficient and effective decision-making and enhance regional economic, environmental, social, and cultural well-being. In turn, regional actions will support national objectives to grow the ocean economy, increase regulatory efficiency and consistency, and reduce adverse impacts to environmentally sensitive areas.

The scope, scale, and content of marine planning will be defined by the regions themselves, to solve problems that regions care about in ways that reflect their unique interests, capacity to participate, and ways of doing business. Marine planning should build on and complement existing programs, partnerships, and initiatives. The intent is to ensure that a region can develop an approach that it determines works best.

This approach balances regional and national interests and recognizes that actions commensurate with regional interests and capacities will provide the most immediate regional benefits. Knowledge and experience will build over time and contribute to achieving national objectives.

- **Support marine planning to advance regionally determined economic, social, environmental, and cultural interests.** States, tribes, and Regional Fishery Management Councils may choose to participate on regional planning bodies established in accordance with the National Ocean Policy Executive Order, this Implementation Plan, and guidance to be released by the National Ocean Council. State, tribal, and Fishery Management Council participation on regional planning bodies is voluntary.

Should all States within a region choose not to participate in a regional planning body within their region, a regional planning body will not be established. Instead, Federal agencies will identify and address priority science, information, and ocean management issues associated with marine planning as described in the Executive Order. In doing so, Federal agencies will coordinate with non-Federal partners and authorities, including States, federally-recognized tribes and Fishery Management Councils, and stakeholders, to ensure that Federal actions support and advance both regional and national objectives.

Marine plans produced by regional planning bodies can provide information about specific issues, resources, or areas of interest to better inform existing management measures. Or, they can describe future desired conditions and provide information and guidance that supports regional action moving forward. Each region has flexibility to build the elements of its plans over time in response to what the region wants to accomplish, the resources available to do the work, and the time it will take to learn what works best in that region. Examples of potential focus areas for marine planning could include, but are not limited to:

- Developing information that facilitates more effective review and permitting among State, Federal, and tribal authorities for a specific class of activity such as offshore energy infrastructure;

- Characterizing environmental conditions and current and anticipated future uses of marine space to assist in siting offshore renewable energy;

- Developing and implementing a plan to acquire data and information to support more efficient management of activities of particular regional interest, such as remote sensing data to support coastal mapping;

- Identifying a specific geographic area and addressing management challenges that would benefit from multi-government resolution;

- Identifying and developing information that better informs agency or government-to-government consultations under the Endangered Species Act, Marine Mammal Protection Act, and the National Environmental Policy Act that apply to offshore development activities important to the region; or

- Developing maps and information that inform effective co-location of multiple existing and new ocean uses, such as commercial fishing, military training, and new energy infrastructure development.

Robust stakeholder engagement and public participation are essential to ensure that actions are based on a full understanding of the range of interests and interactions that occur in each region. Consultation with scientists, technical experts, the business community, and those with traditional knowledge is a foundation of marine planning.

Regional planning bodies are not regulatory bodies and have no independent legal authority to regulate or otherwise direct Federal, State, tribal, or local government actions. All activities will continue to be regulated under existing authorities. For example, commercial and recreational fishing will continue to be managed exclusively by the relevant State and Federal fisheries managers and Regional Fishery Management Councils or Commissions.

As an initial action, the National Ocean Council will provide additional guidance to support marine planning in the regions that choose to move forward through regional planning bodies as described above.

photo: NOAA

VI. Science and Information

Scientific and technological advances allow us to better understand our world. Building our knowledge allows us to respond more appropriately to new challenges, adapt to changing conditions, and take advantage of emerging opportunities for the benefit of our Nation. Strong science, technology, and engineering capabilities and informed people and communities are the foundation for improving our understanding of the marine environment—from the coasts to the deep sea—and informing our decisions about how best to manage the activities that affect the valuable and multiple resources the marine environment provides.

Sustained scientific research and innovative technologies give us the high-quality information we need to maintain or restore ocean resources, guide development and investment opportunities, safeguard lives and property from marine hazards, enhance national security, prepare for and respond to the impacts of climate change and ocean acidification, improve public health, and protect ocean resources. Advancing our scientific, technological, and engineering capabilities also increases the Nation's competitiveness and helps spur the innovation that drives our economy and improves our quality of life. Ultimately, success in improving the ways we use and manage ocean resources depends on building broad public understanding and recognition of the importance of the ocean, coasts, and Great Lakes to our daily lives and the long-term welfare of our Nation.

The actions in this section will engage partners and stakeholders to provide significant, long-term commitments of scientific, technological, and educational support to address existing priorities and apply new knowledge to improve our approaches to management and inform our responsible pursuit of opportunities. Discoveries and technological advances will provide data and information to improve decision-making and enhance the effectiveness of management actions. A focus on fundamental research and exploration will ensure continued advances in basic scientific understanding. An informed society will enable innovative and effective entrepreneurship and stewardship. Collectively, these actions will provide information and capabilities needed to support economies, improve human well-being, enhance environmental health, and increase safety and security.

Enhancing Our Understanding of Ocean and Coastal Systems

For the United States to continue to be a global leader in understanding and acting on the connections between our well-being and the health of the natural environment, we need to continue exploring and expanding our knowledge of the ocean, our coasts, and the Great Lakes. Management and policy decisions must be based in the context sound science provides, through the integration of natural and social science data, information, and knowledge. National Ocean Policy actions will contribute to high-quality science and ensure that information based on that science is made available to guide decisions and actions. Insight gained from scientific research, advances in observations, and innovative technologies will further enable evaluation of trade-offs between alternative management scenarios, enhance our ability to balance competing demands on ecosystems, and strengthen our Nation's economic and scientific competiveness. At the same time, increasing understanding of the ocean, coasts, and Great Lakes among our people and communities will empower better-informed public stewardship of ocean resources.

- **Advance fundamental scientific knowledge through exploration and research.** Through Federal research and exploration activities and partnerships with non-governmental organizations, new ocean discoveries will expand our knowledge and understanding of oceanic and Great Lakes biodiversity, biogeochemical processes, ecosystem services, and climate interactions. Agencies will use the Ocean Research Priorities Plan, a document built with input from the ocean science and technology community, as a reference in determining research directions. They will conduct expeditions in poorly known or unknown regions of the ocean and Great Lakes. They will also work to incorporate natural, social, and behavioral-science information in decision-support tools, which will enable Federal, State, tribal, regional, and local authorities to manage ocean, coastal, and Great Lakes resources more efficiently and effectively.

- **Advance technologies to explore and better understand the complexities of land, ocean, atmosphere, ice, biological, and social interactions on a global scale.** Environmental observation provides the basis for informing decision-making. New technologies, including improved remote sensing systems, and the coordination among agencies needed to develop and implement them, are critical to improving our understanding of the underlying physical and ecological processes driving the ocean, coasts, and Great Lakes, as well as to identifying more efficient means of monitoring these ecosystems. Federal agencies will evaluate how to most effectively integrate observational data, test and develop ocean sensors and communication standards, and implement data and modeling techniques to support a global observational capability to show how observed variables change over time.

- **Increase ocean and coastal literacy.** Increased public understanding of ocean and coastal science and the importance of the ocean in how our planet functions will empower people and communities to be better stewards of ocean resources and increase awareness of opportunities related to these resources. It will also increase interest in activities to address the issues facing the ocean, our coasts, and the Great Lakes. Agencies will contribute to opportunities for systematic inclusion of ocean topics and concepts into mainstream K-12 and informal education systems. Agencies will also develop content that incorporates the latest ocean science for use in schools, aquariums, science centers, National Parks, and other institutions, and conduct demonstration projects that deliver ocean observing data for schools and other educational opportunities.

Strengthening Our Ability to Acquire Marine Data and Provide Information

Vital to ocean and coastal research and management in the United States is the availability of modern ships, undersea vehicles, moorings, satellites, laboratories, instruments, and observing systems. Our ability to send sensors and scientists to sea through these facilities and infrastructure provides critical information for protecting human lives and property from marine hazards, enhancing safety and security, understanding and projecting global climate change and ocean acidification, improving ocean health, and providing for the protection, sustainable use, and enjoyment of ocean resources. Improved science and technology will help the scientific community forecast changes with greater certainty and provide guidance for communities, resource managers, and commercial interests alike.

- **Assess the status of the Federal Oceanographic Fleet to inform future planning and to ensure a more efficient interagency approach to managing the Fleet.** The Federal Oceanographic Fleet is a critical national infrastructure that supports Federal agency and academic oceanographic operations, surveys, and research across a broad spectrum of needs. Ships provide access to the sea and Great Lakes and enable us to gather critical information that supports our responsible use and management of marine resources. Federal agencies will use the inventory and status report of the Federal Oceanographic Fleet to identify its capacity to support a range of requirements nationwide, including in the Arctic, such as data collection and research, weather and climate, ocean mapping, and the understanding of ocean and seafloor physical, chemical, geological, and biological processes.

- **Advance and sustain ocean, coastal, and Great Lakes observing system infrastructure to support a variety of users.** The hard infrastructure of our ocean observing systems, which include various sensors and instruments affixed to buoys, gliders, piers, sea walls, and other platforms (e.g., satellites), form the foundation for a national, integrated observing system which yields real-time information about the marine environment, including meteorological, oceanographic, and ecological conditions. Such information is of value to many different users—from commercial and recreational interests to government and academia—on a daily basis for multiple purposes. Federal agencies will work to advance and sustain the infrastructure of ocean observing systems, such as the Integrated Ocean Observing System and the Ocean Observatories Initiative, to ensure their capability, reliability, and longevity in providing valuable data and information to a growing community of users. Federal agencies will also develop a national ocean observation and monitoring plan to address new autonomous underwater vehicle technologies and sustained monitoring of the water column.

- **Develop an integrated ocean and coastal data and information management system to support real-time observations.** Agencies will coordinate to develop a nationally integrated information management system for our ocean observing systems. We need this system—with supporting interagency data management policies—to realize the full potential and benefits of the Nation's investment in ocean, coastal, and Great Lakes observing systems. This effort will provide easy access to relevant ocean observing data and information for research, planning, and decision support, and will be closely linked with the national marine planning data portal (ocean.data.gov) and other ocean and coastal data portals and services.

- **Implement a distributed biological observatory in the Arctic to monitor changes and improve our understanding of their socioeconomic and ecosystem impacts.** The effects of Arctic changes and human activity on ecosystems and Alaskans who depend on them are poorly understood. Continued observations are needed to form a basis of understanding of the changing processes in the Arctic region. Agencies will continue to develop and deploy a distributed biological observatory, or an array of sites for consistent monitoring of biophysical responses in the Arctic marine environment, as a component of the integrated Arctic Observing Network. Regional collaboration and partnerships will increase our capacity to monitor and assess changing environmental conditions and support improved management of Arctic coastal and ocean resources.

Improving Science-based Products and Services for Informed Decision-Making

High-quality science and information are the foundation for the development of new and improved products and services, including decision-support tools and information displays, which can help inform the decisions at all levels of government working to protect and sustain our economy and environment. Federal agencies will pursue the following actions to enable all interests and decision-makers to make the best-informed decisions possible.

- **Improve the science framework to support decision-making.** Implementing ecosystem-based management will require support from all available scientific tools and methods (e.g., observing, monitoring, synthesizing, hypothesis testing, modeling, predicting, and reporting). Agencies will identify gaps in the basic natural and socioeconomic data needed to advance development and practice of ecosystem-based management, and develop plans to fill them, and engage partners and stakeholders in development of guidelines and best practices.

- **Provide the high-quality data and tools necessary to support science-based decision-making and ecosystem-based management.** Robust decision-support tools and processes will provide information derived from natural and social sciences and traditional knowledge to support timely and effective decision-making. To the degree practicable, these tools and processes will take advantage of and build upon Federal, State, tribal, regional, and local data portals and regional data sharing systems, and be coordinated with other Federal data policies and initiatives.

- **Develop and share decision-support tools to identify coastal land protection and restoration priorities.** Developing and sharing decision-support tools will promote better coordination between Federal agencies and local, State, regional, and tribal entities in identifying protection and restoration priorities across the coastal landscape. As a pilot project, Federal agencies will complete the initial build-out of a Chesapeake Bay decision-support tool system and institute collaborative partnerships within the Bay to support coastal land conservation and restoration planning.

photo: Ernest Koe

VII. Conclusion

This Implementation Plan identifies practical, efficient, and responsible actions that Federal agencies will take to support healthy, productive, and resilient ocean, coastal, and Great Lakes waters, thriving coastal communities, and a robust, safe, and secure marine economy. The Plan will strengthen and build on existing relationships, help forge new partnerships, and enable broad participation from stakeholders and the public in decisions that impact the oceans, coasts, and Great Lakes. Fundamentally, it will provide the science and tools our Nation needs to sustain and enhance the quality of life for all Americans.

Endnotes

1. Throughout the Implementation Plan, the term "State" includes Puerto Rico, the U.S. Virgin Islands, Guam, the Commonwealth of the Northern Mariana Islands, and American Samoa.

2. "Tribes" refers to Federally-recognized tribes.

3. The Implementation Plan is intended to be read in conjunction with Executive Order 13547 and the Final Recommendations of the Interagency Ocean Policy Task Force, July 19, 2010.

4. NOAA Coastal Services Center. 2012. NOAA Report on the Ocean and Great Lakes Economy of the United States. http://www.csc.noaa.gov/digitalcoast/_/pdf/econreport.pdf (accessed February 2013).

5. NOAA National Ocean Service, Special Projects Division, Spatial Trends in Coastal Socioeconomics (STICS). 2013. http://coastalsocioeconomics.noaa.gov/ (accessed February 2013), and NOAA Office of Program Planning and Integration. 2013. The Ocean and Coastal Economy: A Summary of Statistics. http://www.ppi.noaa.gov/wp-content/uploads/coastal-economy-pocket-guide-2011-03-27.pdf (accessed February 2013).

6. Ibid.

7. Lovell, Sabrina, and Susan Stone. 2005. The Economic Impacts of Aquatic Invasive Species: A Review of the Literature. U.S. EPA, National Center for Environmental Economics, Washington, DC. http://yosemite.epa.gov/ee/epa/eed.nsf/ffb05b5f4a2cf40985256d2d00740681/0ad7644c390503e 385256f8900633987/$file/2005-02.pdf (accessed February 2013).

8. Cesar, H., Burke, L., and Pet-Soede, L., 2003. The Economics of Worldwide Coral Reef Degradation. Cesar Environmental Economics Consulting (CEEC), Arnhem, The Netherlands. http://assets.panda.org/downloads/cesardegradationreport100203.pdf (accessed February 2013).

9. NOAA National Ocean Service, Office of Response and Restoration. Marine Debris Information: Economic Impacts. http://marinedebris.noaa.gov/marinedebris101/economics.html (accessed February 2013).

10. Lowther, Alan, ed. 2012. Fisheries of the United States - 2011. NOAA National Marine Fisheries Service, Office of Science and Technology, Fisheries Statistics Division, Washington, DC. p. 93. http://www.st.nmfs.noaa.gov/st1/fus/fus11/FUS_2011.pdf (accessed February 2013). See also NOAA's aquaculture fact sheet located at http://www.nmfs.noaa.gov/aquaculture/faqs/faq_aq_101.html (accessed February 2013).

11. Ibid, p. 58.

12. Marine Fisheries Advisory Committee. 2007. Vision 2020: The Future of U.S. Marine Fisheries. U.S. Department of Commerce, Washington, DC. http://www.nmfs.noaa.gov/ocs/documents/Vision_2020_FINAL-1.pdf (accessed February 2013).

13. NOAA Coastal Services Center. Digital Coast. 2013. Economics: National Ocean Watch (ENOW). http://www.csc.noaa.gov/digitalcoast/data/enow (accessed February 2013).

14. NOAA National Ocean Service, Special Projects Division, State of the Coast. 2012. Economy: Ports – Crucial Coastal Infrastructure. http://stateofthecoast.noaa.gov/ports/welcome.html (accessed February 2013).

15. NOAA National Marine Fisheries Service, Office of Science and Technology. 2013. Fisheries Economics of the U.S. (2011).
http://www.st.nmfs.noaa.gov/economics/publications/feus/fisheries_economics_2011 (accessed March 2013).

16. U.S. Department of the Interior. 2012. The Department of the Interior's Economic Contributions, FY 2011.
http://www.doi.gov/americasgreatoutdoors/loader.cfm?csModule=security/getfile&pageid=308931 (accessed February 2013).

17. U.S. Department of Energy, Office of Energy Efficiency and Renewable Energy, Wind & Water Power Program, and U.S. Department of the Interior, Bureau of Ocean Energy Management, Regulation, and Enforcement. 2011. A National Offshore Wind Strategy: Creating an Offshore Wind Energy Industry in the United States.
http://www1.eere.energy.gov/wind/pdfs/national_offshore_wind_strategy.pdf (accessed February 2013).

18. Restore America's Estuaries. 2011. Jobs & Dollars: Big Returns from Coastal Habitat Restoration. http://www.estuaries.org/images/81103-RAE_17_FINAL_web.pdf (accessed February 2013), and Edwards, P.E.T., Sutton-Grier, A.E., and Coyle, G.E. 2012. Investing in nature: Restoring coastal habitat, blue infrastructure, and green job creation. Marine Policy. Vol. 38. pp. 65-71.

19. Lowther, Alan, ed. 2012. Fisheries of the United States - 2011. NOAA National Marine Fisheries Service, Office of Science and Technology, Fisheries Statistics Division, Washington, DC. p. 20. http://www.st.nmfs.noaa.gov/st1/fus/fus11/FUS_2011.pdf (accessed February 2013).

20. Knapp, G. 2008. Potential Economic Impacts of U.S. Offshore Aquaculture. In Rubino, M., ed. 2008. Offshore Aquaculture in the United States: Economic Considerations, Implications & Opportunities. Silver Spring, MD: NOAA National Marine Fisheries Service. NOAA Technical Memorandum NMFS F/SPO-103. 263 pages.
http://www.nmfs.noaa.gov/aquaculture/docs/economics_report/econ_report_all.pdf (accessed February 2013).

21. Pimentel, D., Zuniga, R., and Morrison, D. 2005. Update on the environmental and economic costs associated with alien-invasive species in the United States. Ecological Economics. No. 52. pp. 273-288.

22. Stedman, S. and Dahl, T. 2008. Status and Trends of Wetlands in the Coastal Watersheds of the Eastern United States 1998-2004. NOAA, National Marine Fisheries Service and U.S. Department of the Interior, Fish and Wildlife Service. http://www.fws.gov/wetlands/Documents/Status-and-Trends-of-Wetlands-in-the-Coastal-Watersheds-of-the-Eastern-United-States-1998-to-2004.pdf (accessed February 2013).

23. Cesar, H., Burke, L., and Pet-Soede, L., 2003. The Economics of Worldwide Coral Reef Degradation. Cesar Environmental Economics Consulting (CEEC), Arnhem, The Netherlands. http://assets.panda.org/downloads/cesardegradationreport100203.pdf (accessed February 2013), and Bishop, R.C., et al. 2011. Total Economic Value for Protecting and Restoring Hawaiian Coral Reef Ecosystems: Final Report. Silver Spring, MD: NOAA Office of National Marine Sanctuaries, Office of Response and Restoration, and Coral Reef Conservation Program. NOAA Technical Memorandum CRCP 16. 406 pages. http://coralreef.noaa.gov/aboutcrcp/news/featuredstories/oct11/hi_value/resources/protecting_restoring_hawaiian_cre.pdf (accessed February 2013).

24. National Invasive Species Council. 2008. 2008-2012 National Invasive Species Management Plan. p. 4. http://www.invasivespeciesinfo.gov/council/mp2008.pdf (accessed February 2013); Pimentel, D., Zuniga, R., and Morrison, D. 2005. Update on the environmental and economic costs associated with alien-invasive species in the United States. Ecological Economics. No. 52. pp. 278-279; and McCormick, F.H., Contreras, G.C., and Johnson, S.L. 2010. Effects of nonindigenous invasive species on water quality and quantity. In Dix, M. and Britton, K., eds. 2010. A dynamic invasive species research vision: Opportunities and priorities 2009-29. Gen. Tech. Rep. WO-79/83. Washington, DC: U.S. Department of Agriculture, Forest Service, Research and Development. pp. 111-120. http://www.fs.fed.us/research/docs/invasive-species/gtr_wo79_83.pdf (accessed February 2013).

25. NOAA National Ocean Service, Special Projects Division, Spatial Trends in Coastal Socioeconomics (STICS). 2013. Demographic Trends Database: 1970-2010. http://coastalsocioeconomics.noaa.gov/ (accessed February 2013); NOAA National Ocean Service, Special Projects Division. State of the Coast. 2011. Climate: Vulnerability of Our Nation's Coasts to Sea Level Rise. http://stateofthecoast.noaa.gov/vulnerability/welcome.html (accessed February 2013); and U.S. Global Change Research Program. 2009. Global Climate Change Impacts in the United States. http://www.globalchange.gov/publications/reports/scientific-assessments/us-impacts (accessed February 2013).

www.ingramcontent.com/pod-product-compliance
Lightning Source LLC
Chambersburg PA
CBHW081407170526
45166CB00010B/3245